Roofing Business and Leadership

An Overview of the Roofing Industry

Developed with funding provided by
The Roofing Alliance

© 2025 Nieri Department of Construction and Real Estate Development
and the Roofing Alliance

ISBN 978-1-63804-186-3
eISBN 978-1-63804-187-0

Distributed by Clemson University Press

Roofing Business and Leadership

An Overview of the Roofing Industry

Dr. Dhaval Gajjar

Dr. Jason Lucas

Editors

Dhaval Gajjar, Ph.D., FMP, SFP

Dr. Dhaval Gajjar is an Associate Professor at Clemson University's Department of Construction and Real Estate Development in the College of Architecture, Arts and Construction. Dr. Gajjar has significant construction industry experience working as a Project Manager for both the general contractor and an owner organization responsible for large remodel, renovation, and IT projects. Dr. Gajjar also has 10+ years of significant research experience related to workforce development, education and training, project delivery and performance measurement. He has authored over twenty (20+) refereed journal publications, one book chapter, and over fifteen (15+) conference presentations disseminating the research results. He has also conducted over fifty (50) industry presentations educating industry professionals on using the research tools. He is also a recipient of Bennett Award and is a certified Facility Management Professional (FMP) and Sustainable Facility Professional (SFP).

Jason Lucas, Ph.D.

Dr. Jason Lucas is an Associate Professor in the Department of Construction and Real Estate Development at Clemson University. He holds a Bachelor of Architecture degree from the New Jersey Institute of Technology and a master's in Building Construction Science and Management and a PhD in Environmental Design and Planning from Virginia Tech. Dr. Lucas has conducted research in workforce development, online education, and the use of technology in education. He has over 10 years of research and teaching experience at Clemson University and has published over twenty (20+) journal articles and over thirty (30+) conference presentations.

Acknowledgements

Assistant Editor:

Gopika Viswanathan

Gopika Viswanathan is a graduate student in the Masters of Construction Science and Management program. She also works as a Research and Teaching Assistant at Clemson University.

Layout Designers:

Danett Vargas Sanchez

Danett Vargas Sanchez is an undergraduate student double majoring in Art and Communication at Clemson University. She is part of the Clemson University Honors College and has four years of graphic design and fine art experience.

Amelia Lyles

Amelia Lyles is an undergraduate student majoring in Graphic Communications at Clemson University. She also works as a Teacher Assistant for Clemson and has two years of graphic design experience.

Technical Reviewers:

William Good, Roofing Alliance

Roofing Alliance Staff:

Alison L. LaValley, CAE

Jessica Priske, Director

Nicole Christodoulou, Manager

Maggie Kosinski, Manager

Acknowledgements

Contributors:

William Good, The Roofing Alliance

Steve Little, KPost Roofing & Waterproofing

Candace Klein, Klein Contracting

Heidi J. Ellsworth, RoofersCoffeeShop

Karen L. Edwards, RoofersCoffeeShop

Mark Standifer

Tom Walker, ABC Supply Co., Inc.

Chris Huettig, KARNAK

Greg Bloom, Beacon Building Products

Erik Zadrozny, Beacon Building Products

Preface

This series of manuals on the roofing industry is made possible through the support of the Roofing Alliance and numerous industry professionals. The manuals were developed from lectures given by industry experts and are meant as an introduction to the topics. It is not intended to include the entirety of the industry and the information about all types of roofing products or systems. For a more complete discussion of roofing materials and installation methods, the reader is encourage to see the NRCA Roofing Manual, available from the National Roofing Contractors Association (www.nrca.net).

The information contained has been reviewed for technical accuracy and clarity at the date of its publication. Codes and practices change over time, so the editors intend to periodically review, revise, and publish future editions of these manuals to reflect those changes.

The roofing manual series can be used independently to provide an overview of different parts of the roofing industry but also serve as a complimentary summary of the knowledge presented in the Clemson Online Professional Development series of courses that have been created with the support of the Roofing Alliance:

- Roofing Fundamentals
- Roofing Management
- Roofing Business and Leadership

The content is broken up into three topic areas to allow tailored focus of employees and business leaders to focus on the area most relevant for them. Roofing Fundamentals provides a general overview of the industry, the products and services available, and focuses on developing an understanding of systems and terminology. Roofing Management covers topics related to a project, including codes, scheduling, field crew management, quality control, risk management, and site logistics. Finally, Roofing Business and Leadership delves into leadership strategies, sales, marketing, and various aspects of owning a business within the roofing industry.

This project would not have been possible without the numerous industry supporters who have donated time and content in support of creating these educational resources.

Table of Contents

Table of Contents

CHAPTER
01

Roofing Contractors Business Overview

With content donated by
William Good
The Roofing Alliance

1.1 Roofing Contractors and General Contractors

Roofing contractors and general contractors both operate in the construction industry, and there are some obvious differences between the two.

As specialty, or subcontractors, roofing contractors only focus on the work described in their contract – either with an owner or a general contractor. General contractors, on the other hand, must focus on the overall scope of the project they are contracted to perform, working with multiple subcontractors that include the roofing contractor.

This distinction creates a number of differences in the way those two types of companies operate. This chapter will focus on the operation of a roofing company.

1.2 Roofing Market background

Most roofing companies are privately owned with a large majority of them being family businesses. According to a study conducted by faculty members at Arizona State University, there are about 50,000 roofing companies in the U.S., and the majority of them operate in the residential market with revenues of less than a million dollars a year. Often, the owner of the company works on the roof, does estimating and keeps the books.

The roofing industry has two distinct markets: residential and commercial/industrial. People in the roofing industry tend to distinguish between steep-slope roofs and low-slope roofs. While steep-slope roofs are found predominantly on houses, they are also found on such buildings as churches and fast food restaurants.

Low-slope roofs make up the majority of the industry by total sales, as you would expect. And the low-slope roofing market requires more sophisticated contractors than the steep-slope market, for reasons that will be discussed below.

Due to a combination of consolidation and organic growth, there are now some very large low-slope roofing companies operating in the U.S.; in fact, two of them have sales in excess of a billion dollars annually.

However, the market for both low-slope and steep-slope roofing contractors remains very fragmented. With a total market of more than $50 billion in the U.S., the top three companies account for less than 5% of market share combined.

1.3 Selling in the Steep-Slope (Residential) Market

Selling in the residential roofing market involves:

1. Knowing who the customer is.

Because some 75% of residential roofing work involves repairing or replacing an existing roof, the customer will usually be the homeowner. For new construction, the customer will usually be the builder, developer and/or general contractor.

Selling to homeowners typically involves responding quickly to phone calls or leads; setting up a meeting with the homeowner(s) and making a formal sales presentation. With the emergence of new technology, contractors are now able to show homeowners aerial photographs of their roofs and virtual examples of what their roof will look like with different types of roof systems.

If there is damage to the roof resulting from a weather event, the insurance adjuster may become a customer, too.

2. Knowing what you are selling.

Studies have shown that roof purchases are made based on a variety of factors, including:

- The price. The same studies show that price is seldom the most important factor, however.
- Service and reliability. Will the contractor be there if something goes wrong?
- Reliability. What is the contractor's reputation?
- Communication. Contractors often fail to notify the homeowner if there are going to be delays due to weather or other issues.
- Cleanliness. Will the landscaping be affected by the roofing work?
- Warranties. Most manufacturers today offer long-term warranties for their products.

3. Knowing the value proposition.

Some homeowners want their roof repaired as quickly as possible. Others want the lowest possible price. Others want to know that the contractor will clean up after the work is done. And still others want to know that their new roof will be made of recycled products and help reduce their energy costs.

1.4 Selling in the Commercial/Industrial (Low-Slope) Market

The commercial/industrial roofing market is generally much more complex than the residential market. And as with residential roofing, it's important to understand how the market operates in order to successfully sell roofing projects. For example, selling in the low-slope roofing market requires.

1. Knowing who the customer is.

This is often not clear-cut, and in many cases involves more than a single individual or entity. Consider:

- Ultimately, the final decision rests with the building owner. But that owner could be a government agency, a school board, a Real Estate Investment Trust (REIT), a private equity firm, an individual building owner or a large, multi-building owning corporation.

- To further complicate things, a building owner may involve an architect or designer for a new or significant project, and that person may have preferences within the contractor community. The owner might also retain a roof consultant, whose job is to oversee the entire roofing project, including recommending the contractor to use. Most roofing material manufacturers have a sales force to identify work for their contractor customers, and many roofing material distributors will have a say in what materials are available, or are discounted – and may also have relationships with local building owners.

- A big part of the commercial/industrial market in the past few years has been roofing huge warehouses, distribution centers and computer cloud storage facilities. Often, the owners of these facilities of large national corporations that choose contractors based on scalability and look for multi-building discounts. Their local facility managers may – or may not—play a role in the final decision-making..

2. Understanding what you are selling.

Among the things most customers prioritize are:

- Brand and reputation of the contractor. Because so many others may influence the ultimate buying decision, being known as a good, reputable contractor generally leads to more opportunities to bid and sell work.

- Price. Most customers expect a reasonable price range for their project; many studies, however, show that price is often not the only factor in a final decision. Instead, customers want to know they are getting the best value, which is often not the lowest price.

- Service. Many roofing contractors have separate maintenance and service teams to handle ongoing maintenance and minor repairs. Providing these services – often in the form of long-term maintenance agreements – is not only valuable to the customer but gives the contractor an advantage when a roof replacement is required. In addition, service and maintenance work generally produces higher profit margins than roof installation.

- Warranties. In many cases, the roofing material manufacturer will offer a long-term product warranty and the contractor will offer a warranty for workmanship. Some material warranties are only offered when contractors who are "licensed" or "approved" by the manufacturer to do the installation. It's important to note that warranties also serve to limit the liability of the companies that offer them, and most warranties have a number of exclusions.

- Communication. NRCA conducted a survey of large, national building owners asking them the most important criteria for selecting a roofing contractor. The top-rated answer was communication with the building owner. Many respondents had issues with phone calls not being returned and with the contractor failing to notify the owner of schedule changes or other job-related issues.

3. Understanding the value proposition.

As is the case in the residential market, the value proposition offered by the contractor will vary according to the customer's priorities. For example:

- Some owners will value establishing a long-term relationship with the contractor. Those owners include colleges and universities, hospitals and manufacturing companies that demand a safe jobsite.

- Other owners – such as a REIT, or a strip mall owner – may only be interested in short-term ownership and value price and speed over long-term capabilities.

- Increasingly, building owners want their buildings to be sustainable and environmentally friendly. They likely will want to know about opportunities for energy conservation, using recycled materials and perhaps even installing vegetative or solar roof assemblies. The best roofing contractors will be able to engage fully in those conversations.

The point is: Selling into either market requires a roofing contractor to think about who the customer is, who the decision-making influencers are and what the value is that they are offering.

1.5 Other Issues in Operating a Roofing Company

Running a roofing company demands that the owner wear any number of hats: he or she must be able to read plans and specifications, understand finances, read and fully understand contracts, act as a human resources director, develop a marketing plan and understand the laws and regulations that apply to the business. While all of these are important, a few bear further discussion.

The issue of labor. Of the estimated 200,000 field workers in the roofing industry, only about 22,000 belong to the International Roofers Union. That means almost 90% of the workforce needs to be recruited and trained by their employers rather than through apprenticeship programs managed under labor-management agreements.

What makes matters worse is that the industry – like all industries in construction – is facing an acute shortage of qualified workers.

One result of all of this is that contractors have increasingly been using "independent contractors" instead of full-time employees to perform some or all of their roofing work. While this may appear to be an easy solution, it still presents problems; for example, how the workers are trained and supervised is subject to labor laws and regulations, and the contractor still retains liability for those workers if they are injured on the job.

Building codes. There are model building codes developed through a consensus-based process that are meant to apply to virtually all aspects of building construction. These codes are typically – but not automatically – adopted by code authorities at the state and/or local levels. It is critical for roofing contractors to know and understand how the codes will be applied and enforced in each jurisdiction where they do work. For the roofing industry, the most important codes pertain to wind and fire resistance – the sources of the biggest insurance claims.

Worker training. In addition to the labor issues noted above, one very crucial issue is how workers are trained. There is a wide variety or products used in the industry today – materials ranging from TPO to built-up roofs to metal panels – and each requires a special set of skills to be installed properly. Some manufacturers will assist in conducting training, but it is virtually impossible to expect every worker to be able to install every roofing system on the market.

Regulatory compliance. The roofing industry, like all industries, is subject to regulation at the federal, state and local levels. The roofing industry gets special attention in many cases, though, because the work is relatively dangerous, because the industry operates a lot of vehicles and because roofing waste accounts for a large percentage of landfill use in the country.

1.6 Typical Career Paths in Roofing

Because roofing companies tend to be smaller than general contracting companies, and because they are exclusively privately owned, there are lots of opportunities for rapid advancement for those looking at career options in the construction industry.

A common entry point for a college graduate would be as an estimator or assistant project manager. Some companies have positions that combine the two – the person in that role will estimate the job, complete the sale and managing it to completion. Next steps on the career ladder would most likely include project manager or estimator/salesperson.

There is also demand in the roofing industry for safety directors, human resource managers and marketing directors, to name a few.

In addition, there are countless opportunities in manufacturing, distribution and consulting. And, because of relatively low barriers to entry, there is always the possibility of starting one's own roofing business.

CHAPTER

02

Mission, Visison, Values, and Goals

With content donated by
Steve Little, KPost Roofing &
Waterproofing
Candace Klein, Klein Contracting

Introduction

Creating a strong business plan, understanding the market and creating marketing and sales strategies, obtaining licensing, funding and insurance and projecting budgets are just the beginning of running a business. The business also needs a mission, vision, values and goals – and people to execute them. Creating a team is a critical next step. This chapter defines mission, vision, values and goals and how to build an effective team to execute them. Additionally, this chapter will touch on challenges to running a roofing business, including crisis management.

Note: there are myriad sources of information on this topic that can be easily accessed online. This is meant just to be a brief overview of common considerations.

2.1 Mission, vision, values and goal

1. Mission and Vision

A mission statement describes an organization's reason for existence and how it aims to serve key stakeholders. A vision statement describes the ideal state that the organization wants to achieve and is aspirational. Both relate to the purpose of the business but show why the business exists (mission) and where it is going (vision). For example, the mission of Toyota in the U.S. is "To make ever-better cars, to build a future where everyone has the freedom to move" and their vision statement is "to be the most successful and respected car company in America."

A primary goal of every company is to generate profits. In order to do that, it is critical to build a team and the team's messaging – both internally and to the company's clients – that is understood by all members of the team. At Toyota US, every employee understands the drive for "freedom to move" as well as the company's vision "to be the most successful and respected care company in America."

Mission and vision statements serve as the foundation for an organization's strategic plan. Without knowing who the company is and what it wants to achieve, it is impossible to create goals and the strategies necessary to achieve those goals.

2. Values

A company's values include the principles that guide and direct the organization and create its culture. The values reflect what the company stands for. They act as a guideline for how employees can make decisions and in turn help build the company's culture. Toyota US's values include the following: "Through our commitment to quality, ceaseless innovation and respect to the planet, we strive to exceed expectations and be rewarded with a smile."

A design engineer, understanding this, may propose less waste in the product. A quality assurance manager may decide that a quality level of 99% is not acceptable. And a Toyota car salesperson may decide to go the extra step to be sure all customers leave with a smile. That is what happens when the whole team understands what a company values, and it creates a plan for the company's goals.

3. Goals

A company's goals are defined, tangible measures for success. They establish accountability. They reflect the "how" in actualizing the mission and measuring progress toward achieving the company's vision.

2.2. Building a Team

There is a substantial difference between hiring people and building a team. Companies have roles they need to fill, such as operations manager, chief financial officer and estimator, but if the hiring process does not consider the company's mission, values and culture, it will not lead to building a team. And while any company can say it has a good culture, creating and maintaining one takes lots of work. Culture, in turn, shapes the brand of the company. Ultimately, company culture and branding will lead to more satisfied customers – and more revenue.

What makes a team great? Great teams act as cohesive units to achieve the company's mission. They have trust, commitment and hold each member accountable. Great teams have the right people in the right roles.

An expert in teambuilding, Patrick Lencioni has written extensively on the topic. Two of his books are *The Ideal Team Player* and *The Five Dysfunctions of a Team*. His books are easily accessible; the two concepts can be summarized as follows:

The Ideal Team Player

Patrick Lencioni contends that the ideal team player have three qualities: they are humble, hungry and smart. There are other skills that are important, he says, but if any of these three qualities are missing, the team can go astray.

1. People who are humble will focus on the greater good rather than their own ego. They take ownership of their mistakes and show appreciation for other team players.

2. People who are hungry seek to do more and to learn more. They are self-motivated and take initiative beyond their own scope of work.

3. People who are smart, as Lencioni defines it, have good emotional intelligence. They ask questions, listen actively and are able to respond appropriately.

Lencioni says that a person missing one of the three qualities can be helped by coaching; a person lacking two or three of the qualities is probably not a good fit for the organization.

And one bad team player can crush the motivation of other team members. If the organization's culture is damaged by a bad team player, ideal team players are likely to begin leaving. And the ideal team players are the ones who will give life to the company's mission and vision.

And one bad team player can crush the motivation of other team members. If the organization's culture is damaged by a bad team player, ideal team players are likely to begin leaving. And the ideal team players are the ones who will give life to the company's mission and vision.

The Five Dysfunctions of a Team

Lencioni says there are five characteristics of a dysfunctional team. They are:

1. Lack of trust. A team member who is unable to show any weakness will not be fully open and transparent with the team. That person will also be too afraid to take ownership of mistakes or be willing to ask for help.

2. Fear of conflict. A lack of trust leads to a fear of conflict. This will preclude team members from engaging in open discussions on important topics

3. Lack of commitment. A fear of conflict leads to a lack of commitment. If any team member hasn't bought into the direction the team is moving (and all decisions around it) he or she is no committed to the solution.

4. Avoidance of accountability.

5. Inattention to results.

The good news is that if a team can define these, the dysfunctions are fixable.

2.3 Challenges in Managing a Roofing Business

Managing a roofing business has two major challenges: you and everything else.

You

Working "in" the business and "on" the business are two entirely different concepts. Working in your business refers to managing the business during day-to-day activities such as procurement, estimating, scheduling, project management and dealing with personnel issues.

Working on your business is investing time today to improve the business tomorrow. Working on the business is how future challenges can be overcome.

Being in the C-Suite (as a senior executive) means you constantly work ON the business while you work IN the business. This is how you overcome challenges and build a path to success. Nature abhors a vacuum. If you don't build your path, someone else will build it for you.

One way to invest in yourself and become a good business leader is to get outside help. That help could come in the form of a mentor, or coach – ideally with knowledge of the roofing industry. Or it could come from being part of a peer group of like-minded business owners.

A good business leader must possess the most essential quality of being able to adapt to changing conditions. A really good leader will read about best management practices, will attend workshops and industry events and will be aware of needs in the local community. And a good leader will always be humble enough to keep learning and growing.

Everything Else

The principal source of challenges is people: not having enough employees, not having employees in the right roles, not having the right fit of talent and so on. Many challenges that arise from humans working in a business can be dealt with through alignment.

Alignment in this context refers to putting the right people in the right job and being sure they are productive team members. Employee capabilities and interests must align with job satisfaction and productivity. Did you promote someone who was excellent at their job into a management role where they are failing? Do they have the trust to ask for training? When there is a change in responsibilities and/or circumstances, how is the team performing? Finding talent, hiring talent and keeping talent are perpetual issues in any business.

Capital

When we think of capital, we usually think only of financial capital, i.e., equity, debt, investments and working capital. But equally important is human capital, as shown in Figure 2.3.2, which includes intellectual, social and physical talents and skills. These skills must align with the nature of the roofing business. Planning for the qualities you need in your team members and developing an organization chart are essential.

In the roofing industry, common challenges include cash flow, recruiting, competition, weather and material issues. Recently, of course, the industry also had to contend with a global pandemic. If a company doesn't have enough capital – either financial or human capital – and faces a crisis beyond its control (like a pandemic or a recession), it can make it difficult for the organization to survive.

Financial capital includes having enough equity in the business to take risks, having the ability to take on and manage debt, having investments in the business such as equipment and having the working capital necessary to keep everything running smoothly.

Issues surrounding human capital include having the talent to manage the company. This includes talent in estimating, field labor, field management and risk management. If the correct human capital isn't in place, common issues – like over- or under-estimating work, ineffective marketing, poor installation, improper safety and the like – will become big problems.

Every roofing company that has been in business for any length of time has experienced a crisis of some sort. The ability to manage a crisis is something that should be planned for and perpetually worked on. Organizations like the National Roofing Contractors Association (NRCA) hold workshops on topics ranging from current technical issue to risk management to common pitfalls in contracts. Keeping current on issues that can lead to a crisis is the first defense against it.

In addition, a company must have a crisis plan to help navigate it through potential problems.When a crisis arises:

- There must be one spokesperson and one message for each situation. The same spokesperson speaks to employees, emergency personnel, the media, the customer and anyone else involved.

- Demonstrate to all employees that the team comes before any individual. The company must stand together and business must continue as usual.

- It is important to keep employees calm during times of stress.

- Contact strategic partners and clients and fill them in on the company's version of events before they hear other versions.

- Have a checklist of the company's crisis management plan.

A good example of a recent crisis facing the industry was a severe shortage of materials caused in part by the COVID-19 pandemic and in part by ice storms that shut down raw material manufacturing facilities in Texas. As a result, what had been a two-week lead time for getting materials turned into, in some cases, a ten-month lead time and a significant increase in material costs. None of this could be foreseen.

Some Bonus Material from NRCA

The association sponsored a study a few years ago, asking large, national building owners to identify the factors influencing their choice of roofing contractors. The top-rated factor was communication. One owner told the story of how a roofing crew failed to show up on the promised date (due to high winds) without letting the company know. And the company had spent much of the previous day clearing out the parking lot to enable the roofing company to load material to the roof.

Three tools of communication are said to be the most effective:

1. Communicate in a "neuro" way. "Neuromarketing," as it is called, involves sharing information in the simplest form possible so that people can grasp it in a fraction of a second (preferably aided by the use of visuals and examples) rather than assuming everyone can follow complex data and details.

2. Speak the language of the industry. Convey messages using positive words and focus on what is working. For example, instead of saying "The work quality is very poor," try saying "The work quality needs some improvement." Simply switching out a negative word (in this case "poor") for a positive word ("improvement") can have a massive impact on inspiring and empowering people.

3. Story telling. People always associate with stories and are wired to trust people who display vulnerability. Communicating purpose, goals and plans through a relatable story can significantly impact the audience. Displaying that you are part of a community shows that you care about the people around you and automatically gain people's trust. This also helps to explain why customers are often more likely to choose a contractor that works closely with the local community.

Community involvement is often viewed as a great tool to build company culture and to become known in the local community. But community involvement is important for more than just marketing purposes. Every business thrives when it becomes part of a community. It can empower your team members and often can turn into a team-building activity. Community involvement always starts with the business owner, which will help encourage employees to participate.

Creating a great team is crucial to seeing the company's mission and vision put in place. Keeping a sharp eye on financial and human capital is necessary to manage the daily operations of the company. And preparing for a crisis is an excellent way to not only prevent one from occurring but to minimize its damage when it does occur.

CHAPTER
03

Sales and Marketing

With content donated by
Heidi J. Ellsworth, RoofersCoffeeShop,
Karen L. Edwards, RoofersCoffeeShop
Mark Standifer

Introduction

It goes without saying that marketing is paramount to the success of any business. Effective marketing helps to create awareness about the company's brand and services, acquire new customers and engage with and retain current customers. Data collected through market research also helps to develop new products and services tailored to customer demands. Marketing also becomes a platform to showcase what makes a company unique.

3.1 Types of Marketing

Personal branding and marketing. Marketing yourself is essential when you are establishing a new business. And the roofing industry has always been known as one that is built on relationships, so it's important to devote some effort to getting yourself known. That includes some basic steps, such as:

- Having business cards with you and available wherever you are.

- Setting up live meetings with decision-makers. Personal interactions take time and show value.

- Arriving early for appointments. Respect the time your customer or prospect is giving you.

- Following up every meeting with an email or letter summarizing the discussion.

- Joining professional organizations that can help you find new strategic partners and keep you current on industry trends.

Establishing client relationships. Having long-term relationships in the roofing industry is an important part of a company's marketing efforts. Some building owners, for example, want a reputable contractor to call in emergencies. Others want to have the contractor provide ongoing roof maintenance services. When it's time for a new roof, the contractor with that relationship has a clear advantage. Some other techniques for building client relationships include:

- Organizing activities a client might enjoy. This could be as simple as taking a client to dinner, or as involved as getting a client involved in a charitable community event.

- Getting branded company merchandise into the hands of customers. It will keep your name in front of them, and help reinforce your brand.

- Make sure your employees understand they have a responsibility to help market, too. They can do that by wearing branded clothing on the job, but even more important, by acting professionally at all times.

Marketing during a project. When you're executing a project, it's the best possible time to showcase your company. There are any number of ways to do that, including:

- Always attending pre-job or pre-construction meetings in person. Here, you'll review scheduling, safety plans, site logistics and availability of materials. Be prepared with a solid plan as you go into these meetings and be ready to answer any questions.

- Be sure to understand your client's expectations. If there are any changes to the agreed-upon plan, communicate them and discuss their impact on the job.

- Demonstrate the value you place on worker safety. Understand what the client expects; understand what other trades will be doing (on a new construction project) and discuss your own company's safety plan.

- Train your employees on proper installation of the materials, but also stress the importance of having them behave professionally. During a project, they will become a big part of your company's brand.

- Communicate! Building owners often cite lack of communication as the biggest issue they have with roofing contractors.

- Arrange for excellent customer service after the job is completed. You can leave a lasting impression by providing a complete close-out package, setting up future inspections and responding quickly to any issues that may arise.

3.2 Sales and Marketing Cycle

The process of sales and marketing forms a cycle. After a company is started, the first step is attracting potential customers to you website, to talk to you in person or to meet at an event. Some standard tools for attracting customers using your computer involve using blogs, social media and search engine optimization (SOE). Once strangers become your visitors, the next step is to convert them into leads. Once a visitor becomes a lead, then you work to close the sale and turn the lead into a customer. Customers then need to be given a good experience from the beginning to the end of a job. By doing a quality job, and treating the job site respectfully, you will have a delighted customer, who can then help bring in more strangers and potential customers to form a new cycle.

Traditional marketing used to involved using the Yellow Pages, which was the advertising section in telephone directories. Companies would often have company name starting with "A," or "AA" or even "AAA" to appear at the beginning of the Yellow Pages. Happily, in the last decade the industry has made great strides in adopting other marketing strategies.

The fundamentals of marketing, though, remain the same. To develop a marketing plan and strategy for the company, the following questions need to be addressed:

- What is the company's priority?
- What is your market?
- Who is your audience?
- What is the budget?

The first step for a company building a marketing plan is to identify core values and develop a mission statement.

The company's core values must match with the experience delivered to its customers. Satisfied customers will always return, and they will help promote the company through positive word-of-mouth. It is also important to consider the products you are selling and what the right markets are for them. Typically, a company's owner decides what products and services to sell, and the sales team reinforces the goals of the company while promoting those products and services.

3.3 Getting a Lead, Long-Term Sales and Vendor Relations

Marketing in the roofing industry is very different for the residential and commercial/industrial markets. While marketing in the residential sector involves marketing to a homeowner (called Business-to-Consumer, or B2C), marketing in the commercial sector involves marketing to a business owner, or Business-to-Business (B2B) marketing.

In both markets, leads can come from digital platforms, public relations, advertising and in-person events. Converting lead generation to sales works like a funnel.

It is essential for a company today to have a website that is regularly maintained and updated, and that makes it easy for a customer or prospect to get information. Search engine optimization (SEO) is a tool to drive traffic to your website. And fresh, relevant content and even blogs on the website can be used to reach a wider audience. Keywords used on blog posts can bring a person who does an internet search using similar words to the company's website.

The key is to look at traditional marketing methods and reinvent them with new digital tools. Once leads begin to be generated:

- A sales and marketing team is essential to develop long-term relationships that will lead to long-term sales.

- A company's service and maintenance department can play a big role in maintaining positive customer relations by responding quickly and professionally to customer needs.

- Because many referrals come from satisfied customers, it's always important to have happy – and yes, delighted – customers.

In the roofing industry, it is also important to maintain sold relationships with vendors, as they can often help a contractor gain new customers. Many roofing material manufacturers have their own sales and marketing teams, usually selling to companies with lots of square feet of roof area. And those manufactures have their own "licensed" or "approved" contractors that they will provide leads to.

3.4 Building a Marketing Plan

For most companies, the marketing department focuses on creating awareness and building relationships, while the sales department focuses on closing deals and generating revenue. Both play a critical role, and the two departments must always work closely together.

A marketing plan necessarily derives from the company's business plan, or strategic plan. The marketing plan should address such issues as:

- The pipeline of potential customers. Are there too many people at the top of the funnel who are not ready to buy? What's working and what's not working?

- The product offerings. Are we offering the right products and services?

- The competitive environment. What are our competitors doing that we're not? Where are they vulnerable?

- The culture of the company. Are all employees working toward the same outcomes?

- The markets we serve. Are they still the right ones?

- Emerging trends. What market changes are going to affect us?

- People. Do we have the right talent to meet our goals?

- Our vision. Are we doing things today to attain our long-term goals?

Remember that a plan is just that: a plan. Things change, and it is always important to keep evaluating and revising the plan as those changes occur.

Ultimately, the company should identify its core values, define its positioning in the market and then build annual goals. Those goals should be realistic and attainable, so that a good budget can flow from them. The next step will be to identify the marketing tools the company will use – from digital to in-person events. The marketing plan in its final form should include an executive summary, company positioning statement, a summary of markets served, yearly goals, branding strategies, marketing tools, budgets, tasks and timelines

3.5 Building a Brand

Books have been written on branding strategies, but at the end of the day a company's brand must stem from what it wants its customers to feel and what promises it wants to make to them. Think about the difference in brands between Starbuck's and Dunkin' Donuts. One evokes a relaxing workplace experience, while the other evokes a "grab a coffee and go" experience. Both brands work – but they are completely different, and offer different sorts of promises to their customers.

Branding must also be considered when the company develops its communication and messaging strategies. What should a logo convey? What colors and font sizes are appropriate for the message that is being delivered?

Many companies involved their employees in their brand-building by equipping them with company merchandise so they will be brand ambassadors for the company. And it is vital that the whole team understands the company's messaging.

3.6.Budgeting and Market Plan Success

A typical roofing company spends between 3 and 5 percent of its annual revenues on marketing. If a company wants to develop a new offering – such as a service department – or offer a new line of products, marketing expenses may approach 10 percent of revenues in the short term. Many companies also have contingency funds set aside for any new opportunities that may arise.

Many roofing companies rely on outside expert help for developing and executing their marketing plans. They might use, for example, writers, website designers and ongoing internet support. Several manufacturers and distributors also offer funds for cooperative advertising that features their products in addition to the contractor's services. Marketing budgets need to be reviewed at least quarterly and always communicated to the company's leadership.

In summary: Plan and think strategically. Communicate plans and strategies to all employees. Be consistent in messaging. Develop networks and personal relationships to help build the company's brand. And always ask for feedback.

CHAPTER
04

Procurement & Sourcing

With content donated by
Tom Walker
ABC Supply Co., Inc.

4.1 Introduction

The roofing industry is composed of two broad areas: Steep-slope and low-slope roofing. Steep-slope is often considered to be synonymous with residential roofing. However, a minority of residential roofing can also involve low-sloped roofs. Similarly, some commercial roofs involve steep-slope roofs. For the purposes of this discussion, we will think in terms of steep-slope roofs for residential structures and low-slope roofs for commercial structures.

For residential roofing, the distributor is the champion in the supply chain, being responsible for transporting products from one place (usually a warehouse) to another (usually the job site). In commercial roofing, the manufacturer is the supply chain champion.

The difference is due to how each market functions. The distributor is an integral part of a residential roofing project because the distributor purchases bulk quantities of material from the manufacturer and delivers smaller portions of that material to the jobsite on specified days. In other words, the customer requires the distributor's "break bulk" services.

A manufacturer is an important part of a commercial roofing project because larger quantities of materials are typically needed, and customers often require the delivery of several truckloads of materials.

Another difference between residential and commercial markets is the end user. A manufacturer in the residential sector will have multiple customers, including distributors, retailers, large home builders and roofing contractors. A manufacturer in the commercial sector deals primarily with only the roofing contractor. Residential projects have a short sales cycle, while commercial projects often have a long and laborious one.

A house needs a single delivery of materials, whereas a commercial project typically requires multiple scheduled deliveries. Also, storm-created demand is a large part of the business in residential repair and reroofing, but creates only minimal demand in commercial roofing (with obvious and notable exceptions).

4.2 Roofing Supply Chain

The roofing market channel starts with the roofing material manufacturer, as shown in Figure 4.2.1 Then comes the roofing distributor, who takes products in bulk from the manufacturer and unpacks and redistributes them from its warehouse in a "break bulk" concept. Distributors also add value to this channel by helping with logistics, providing technical services and becoming creditors.

Next in the supply chain is the professional roofing contractor who provides labor, installs the roof and understands local building codes and project requirements. The last entity in the chain is the end user or building owner. The supply chain only works, of course, if there is a building owner to provide a roofing project.

Roofing Manufacturer	Roofing Distributor	Roofing Contractor	General Contractor*	Building Owner/ End-user

*only for new commercial projects

Figure 4.2.1 – Roofing Supply Chain
Source: Author

For commercial roofing projects, another key player – the general contractor – is involved only in new construction projects and has been hired by the owner. General contractors are rarely involved in reroofing projects, which represent between 75 and 80 percent of all roofing work.

In about 20% of commercial roofing projects, the manufacturer sells directly to the roofing contractor without the use of a distributor. This is because some smaller manufacturers make niche-oriented products and deliver them directly to contractor customers. Some roofing contractors have also established relationships with larger manufacturers who will sell directly to them.

Money follows the supply chain: the manufacturer invoices the distributor; the distributor invoices the roofing contractor and the roofing contractor invoices the general contractor (in new construction) or the building owner.

4.3 Procurement Considerations

The roofing contractor will also have to procure more than just roofing materials, for example:

Other trade procurement

Other trades. The roofing contractor might have to contract with electrical, plumbing and HVAC contractors in order to properly install the roof. As just one example, a drainage system may need to be installed by a plumbing contractor so that the roofing contractor can make it watertight.

Subcontractors

Often, the roofing contractor will subcontract certain portions of a roofing project. Examples include roofing ballast (with roof vacuums), installing solar panels, installing vegetative roof systems, conducting moisture surveys and installing architectural metal materials when specified.

Ancillary items

These include such things as trash dumpsters and portable toilets which may be needed to support work at the job site.

There are a host of other issues that need to be considered before a roofing project begins, because any of them can add cost to the project. These include:

- Local building codes. The contractor needs to ensure that all local building codes are being met.

- Insurance. In some cases, roofing projects – and reroofing projects in particular – need to meet requirements established by the company that insures the building. For example, Factory Mutual, a large building insurance company, establishes requirements for wind uplift resistance on roofs that are installed on buildings it insures.

- Building use. Different building types have different roofing requirements. For example, it is common to see complex HVAC and other mechanical equipment installed on the roofs of manufacturing plants, but only rarely on warehouses.

- Roof structure and detailing. The roof system must be designed according to the building's structure. For example, a ballasted roof system will not work on a building that is not designed to handle its weight.

- Slope of the roof. Water accumulated on a roof must be drained off efficiently. If the roof does not have sufficient slope to drain the water, tapered insulation can be used to create a slope.

- Rooftop projections. The type and number of rooftop projections, e.g., skylights and vents, must be taken into consideration.

- Wall openings. Openings in the wall play a role in the wind uplift pressure on the roof. For instance, large bay doors can create negative wind uplift pressure on the roof, which must be accounted for in the roof's design.

In addition, there are a number of more general factors that need to be taken into account, including water management, the surrounding topography, live and dead loads on the roof, how the roof will be accessed and maintained and its general aesthetics.

4.4 Procurement Process

From a roofing contractor's point of view, the procurement process for a roofing project occurs in five phases, as shown in Figure 4.4 .1.

| Lead Generation/ Estimating | Quoting | Ordering | Logistics/ Delivery | Invoice/ Collection |

Figure 4.4.1 – Overall Procurement Process
Source: Author

The process begins with procuring leads. For commercial roofing projects, the customer could be a general contractor, building owner, property manager or a representative of the building owner, such as a roof consultant. For residential roofing projects, the customer could be a single homeowner, a homebuilder of a homeowner's association. Once the leads are developed, the next step is estimating the roofing project. Many commercial projects that are going to be bid by several contractors will include a pre-bid notification and a pre-bid meeting that contractors and distributors attend. At the pre-bid meeting, specifications are reviewed and discussions are held with the designer to select suitable materials for the project. It is essential at this stage to ensure that the contractors involved are creditworthy, as material costs alone for commercial roofing projects can be in the tens of thousands of dollars.

The next phase is developing a quote, or proposal. The contractor or distributor reaches out to one or more manufacturers, providing them with the project's details and a list of materials. A negotiation process typically ensues. Following the negotiation, the materials costs and rates are set and the contractor prepares and submits a bid.

Once the job has been awarded, the contractor places an order for the materials; at this point quantities, shipping dates and delivery dates and locations are reviewed.

The final phases are delivery of the material to the distributor, contractor and/or job site, and then invoicing, as discussed above.

CHAPTER
05

The Role of Manufacturing

With content donated by
Chris Huettig
KARNAK

5.1 Introduction

Manufacturing is broadly defined as the process of combining and converting raw materials to form large quantities of valuable products. Manufacturing is an important contributor to the U.S. economy, amounting to nearly 12% of GDP. While many people associate manufacturing with simply the production of goods, many other factors are involved in the manufacturing process. These include the design of products, methods of production, costs of raw materials, machinery required and sales. The manufacturer combines raw materials from various suppliers to create a unique product sold to the industry at a cost greater than its costs of acquiring and producing the materials. Manufacturers are always looking for ways to produce new finished products that benefit the industry or industries they serve.

Manufacturing typically includes three key functions: design, production and sales. These functions are all interdependent and contribute to the smooth operation of the manufacturing business. Each of these functions needs to have its own processes and standard operating procedures. For example, the sourcing of raw materials requires consideration of price, quality and availability.

For the business's overall success, it is also imperative to have the right people. People in leadership positions at manufacturing companies must be capable of making decisions about product selection, the business approach and – always – the mission and vision of the company.

Successful manufacturing companies train their employees at all levels. And of course marketing the company's products is paramount to its success. This involves not only sales, but also packaging, delivering the product to customers and having a sound marketing strategy.

5.2 Bringing Product to Market

The primary role of a manufacturer of roofing materials is to produce quality products that, when assembled with other complementary products, result in a roof assembly that protects the interior of the building from exterior environmental factors. Without good products, a good roof can't be constructed. That's why manufacturers are so important in the roofing industry.

In its simplest form, a manufacturer:

1. Takes an order

2. Manufactures the product

3. Ships the product

However, there are a lot of considerations that go into this process, starting with developing a business case. Is there a need for this product? How will it be useful for building and maintaining roof systems? What are people willing to pay for it?

Once a business case is made, most manufacturers have a "stage gate" approval process that includes the company's department heads to provide information about their involvement in the development of the product. Typically, the departments are as follows:

Marketing

The marketing department typically drives the "stage gate" process by assessing the products, determining how to build a message around it and considering how it should be marketed. Ideally, the marketing team has been working closely with the research and development team during the development of the business case

The marketing department will also work with the other departments to create the finished product. For example, the marketing department will work on package design, sales literature and promotional literature to ensure a successful product introduction.

Research and Development (R&D)

The head of the R&D Department will assess whether the proposed product is plausible. The R&D team then puts together the specifications for the product in detail by considering the characteristics of the raw materials that will be used. The team then creates a prototype to validate that the product can be made ("proof of production"). The prototype is then tested internally to see if it meets the necessary specifications. Often, testing is also conducted externally – in the field – and then test results are submitted for third-party approval by such organizations as Underwriters Laboratories or Factory Mutual. This is done to ensure that the products are safe to use and to validate that the product meets industry standards. Prototypes are commonly tested to ensure they meet building codes.

Finance

The finance team develops the cost to manufacture the product, and its total delivered cost in order to measure its profit margin. The team will then determine the return on investment from the product by assessing factors such as whether new assets or additional people will be needed. The team then can assess whether the product will be financially beneficial to the business.

Technical Services

The technical services team assesses the product by reviewing its list of pre-determined requirements. The team is responsible for creating product documentation, and warranties.

Product documentation typically includes product data sheets (PDS), safety data sheets (SDS), applications guides and design guides. Product data sheets contain a synopsis of product information that contractors, architects and others use to select suitable materials for a project. Safety data sheets contain information regarding the chemical composition of the materials used in the product, personal protective equipment (PPE) that might be needed and overall guidelines on what must be done to ensure that the product is used safely.

Application guides are similar to product data sheets, but they comprehensively cover the methods and tools needed for efficient application of the product.

Design guides include information about how the material must be installed or otherwise be used to maximize the benefits of its design. These guides cover topics like attachment for wind uplift, installation parameters at roof perimeters or eaves, design considerations that need to be taken into account and so on.

The technical services team is responsible for creating warranty documents and an "early warning system" that includes all known scenarios in which the product could fail, ways to determine the reason for failure and how to fix potential issues that may arise.

Additionally, there are post-production technical services that the company must offer. After a product is manufactured, it is essential to train installers on how to use the product. The amount and type of training needed are, of course, dependent on the type of product being offered. The technical services team is also responsible for assisting with the start of the job, working with the crew on the first day of installation to provide guidance and offer suggestions.

If the company is going to offer a warranty for the product, the technical services team will ensure it is properly stored and installed correctly. The team might also be involved in warranty services in the event of a product problem or warranty claim.

Manufacturing

The manufacturing department will assess how the product can best be created on what machinery will be use to make it. The team is also responsible for planning the entire manufacturing process from start to finish, which includes everything from how and when different raw materials arrive to how the final product is packaged. The team is responsible for the scheduling of production and planning, and must consider the best order of product assembly, the number of people needed to work on the product and the time and resources spent on each batch of the product.

Logistics

The logistics team is responsible for ensuring that product shipment regulations are met. For example, if the product is flammable, it must be specially labeled before shipping. The logistics team is also responsible for establishing delivery methods and schedules. Most roofing materials are transported via flatbed trucks; however, some tools or accessories might be needed to be transported more quickly – even overnight. The team will have to determine the most appropriate ways of transportation.

Legal

The legal team will examine all legal issues surrounding sourcing, producing and selling the new product.

The team can be very helpful if a problem or complaint about a product arises, and can also help with issues pertaining to shipping products internationally. The legal team is also responsible for identifying potential liabilities in the event of future failures, as well as types of warranty responsibilities.

Purchasing

The purchasing department will determine how and where raw materials can be sourced. Virtually every roofing product contains several raw materials; it is the purchasing department's responsibility to select raw material suppliers based on the quality of the materials, their availability, their price and the overall qualifications of the company providing them. The purchasing team will always have secondary suppliers as backups in case the primary supplier stops production for any reason. The team has to be able to make quick purchasing decisions, especially when disruptions or changes in the market occur. They are also responsible for managing inventory.

Sales

The sales team will assess and determine the best ways for the product to be sold. The team is responsible for introducing the product to the market. They collaborate with direct users (typically roofing contractors and their applicators) thorough on-site visits, observing their needs and determining what works and what does not.

The sales team also works with distributors to sell the products. Helping distributors understand how the product could be used is very beneficial when the distributor, in turn, sells the product to contractors. The sales team usually works also with end users such as homeowners and building owners, and can also sell through retailers like The Home Depot.

The sales team might also collaborate with architects and roof consultants, as they are critical decision makers on many projects and are often responsible for product selection.

Quality Assurance (QA)

The Quality Assurance department focuses on how the quality of the product can be ensured and how it meets required specifications. The QA team is involved in both the pre-production and post-production stages and is responsible for obtaining the approval of each department in the "stage gate" process.

The team puts together checks and balances at the pre-production stage, so that it can ensure the product can be manufactured according to specification. The team also thoroughly tests products to be sure they meet quality standards and consistency measurements.

Customer Service

The customer service department is responsible for taking orders and scheduling customer shipments and pick-ups. The team undertakes all the daily tasks pertaining to sales and service, and often has the most contact with customers and end users. The team is therefore often the first responder when a problem arises or when situations need to be handled immediately.

5.3 Becoming a Good Manufacturer

In order to be a really good manufacturer, a company must:

- Understand the customer's needs. Market research and customer surveys are common tools businesses use to understand the needs of their target customers.

- Not compete with price alone. Good companies focus on product value rather than on reducing prices to be competitive. Price cutting rarely helps a business, but can threaten its long-term viability as profit margins are reduced. Marketing the brand and emphasizing the value of the products it makes are much more beneficial in the long run.

- Be innovative and constantly look for ideas to improve existing products or manufacturer a new one. Innovation can help manufacturers develop unique products that differentiate them in the marketplace.

- Work as a team to solve problems collaboratively, without blaming others. This kind of teamwork leads to a positive work culture, improves efficiency and productivity and results in better problem-solving.

CHAPTER
06

Role of Distribution

With content donated by
Greg Bloom and Erik Zadrozny
Beacon Building Products

6.1 Introduction

A distributor acts as a "conduit" between a manufacturer and a contractor (or another end-user) to ensure the timely delivery of products on the job site, as depicted in Figure 6.1.1. As discussed in Chapter 4, a distributor procures materials in bulk from the manufacturer, stores and takes inventory at their facility and then distributes products as needed to end-users. Sometimes distributors also ship directly after the purchase.

Manufacturer   **Distributor**   **Installer**

Figure 6.1.1 – Distributor's role as a 'conduit'
Source: Beacon Building Products

Strategic location is important for setting up a distribution facility. The closer the facility is to a dense geographical area with a high demand for these products, the easier it is to avoid transportation problems. Location is also important to collaborate with manufacturers in the area and to attract potential buyers. Distributors today have specialized capabilities to ease the entire procurement process by incorporating new technology, tracking deliveries, adopting better safety measures, using automation and offering lines of credit.

Roofing is a relationship business; most sales between entities are not limited to a one-time purchase. Therefore, establishing working relationships with contractors while working in distribution is essential.

Distributors support contractors in their work by offering credit services, regularly communicating, providing technical assistance and making the supply chain process work as easily as possible.

Distributors often also help contractors with material and product brand selection and bring unique products and equipment to the table. Partnering with distributors can be very beneficial as they function as one-stop shops for a variety of products and materials.

6.2 Product Demand

The two primary markets within the industry are residential and commercial roofing. Both markets require different equipment and technical expertise and involve different selling processes. Distributors sell products and cater to the demands of both markets. Some distributors also sell non-roofing products, such as those outlined in yellow in Figure 6.2.1. These other products could be helpful for contractors handling additional scopes of work on a construction project.

Figure 6.2.1 – Two core markets
Source: Beacon Building Products

Roofing is an industry of high priority and functions steadily even during times of economic crises.

The annual compound growth between 2015 and 2020 was 4.9%. Growth from 2020 to 2025 is expected to be 6.7%. This is indeed promising for the roofing industry and its continued growth.

Major supply chain issues such as those aggravated by the COVID-19 pandemic have increased demand for building materials. The rooing industry still manages to navigate through these troubles in material procurement and continues to grow.

While the physical demand for roofing measured in square feet shows a gradual increase from 2010 to 2030, the monetary market in dollars is growing at a faster rate, particularly from 2020 onwards. This indicates a rising average cost per square foot or a shift towards higher-end roofing materials. The compound annual growth rates (CAGR) for both new roofing and reroofing projects show a marked decrease in physical demand growth post-2020, suggesting a maturing or saturated market, while the monetary growth rates indicate an increase in value, possibly due to escalating material costs or premium product offerings.

6.3 Roles in Distribution

The distributor typically sends trucks loaded with products and materials from its yard or warehouse to the job site. A driver and warehouse worker ensure that the materials are loaded properly. But before the load is sent out, the distributor must ensure that all ticketed products have been loaded and that the quantities sent out are correct. Once at the job site, a delivery helper from the distributor usually works with a roof loader responsible for staging that job. When a roofing contractor gets to the job site, all necessary materials are loaded safely onto the roof.

It is crucial to keep the job site safe. Distributors also distribute the roof load on the roof at pre-identified locations and drop materials to the ground with specialized equipment.

The delivery helper/driver returns with the empty truck to the distributor's yard, and the process is repeated when another order is received. These three stages are displayed in Figure 6.3.1 below.

| Driver + Warehouse Worker | Delivery Helper | Roof Loader |

Figure 6.3.1 – Roles involved in a typical material delivery
Source: Beacon Building Products

Traditional field roles in distribution include:

- Delivery Helper
- Driver
- Warehouse Worker
- Inside Sales Representative
- Outside Sales Representative
- Assistant Branch Manager
- Branch Manager
- District Manager
- Sales Director

Every distributor needs people assigned to these roles to support the business. Some smaller distributors with fewer employees might have some employees assigned to more than one of these roles. The District Manager and Sales Director are responsible for overseeing all branches and the business in general. The District Manager manages the operations side of the business, while the Sales Director manages overall sales. As with all businesses, there are traditional support roles involved with all of these functions.

Those include: the Executive Team (CEO, COO, etc.), Credit, Financial Planning and Analysis (FP&A), Marketing, Safety, Human Resources, Information Technology, Supply Chain Management, Pricing and Accounts Payable/Accounts Receivable. All of these roles, whether field or support, play a critical role in the smooth functioning of the business.

The distribution business is, of course, highly dependent on manufacturers. It is therefore of utmost importance for distributors to form close working relationships with roofing manufacturers.

In 2020, the three largest roofing material manufacturers (GAF, Owens Corning and CertainTeed) supplied 39% of the US roofing market. Carlisle SynTec Systems, Johns Manville and Holcim/Elevate together accounted for another 16% of roofing sales. Figure 6.3.2 is a map showing the location of the manufacturing facilities of these companies. Distributors must be aware of potential manufacturers from whom materials could be sourced.

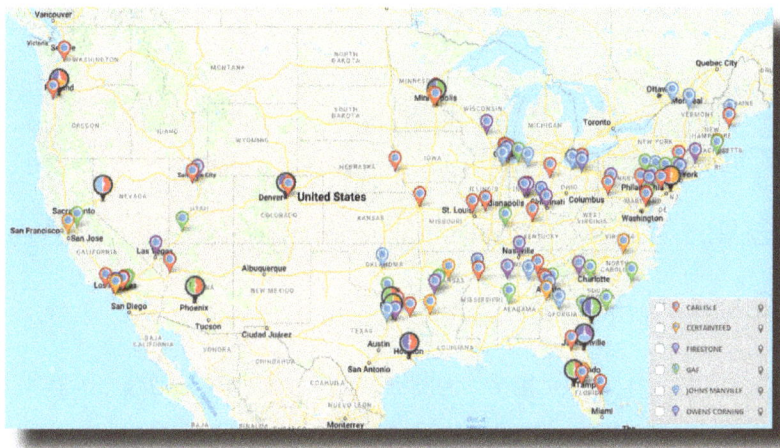

Figure 6.3.2 – Manufacturing locations
Source: Beacon Building Products

The location of manufacturing facilities significantly impacts the locations of distribution yards and branches. A strategic location helps to reduce travel time and help with faster procurement of materials, making the business more resilient in the event of supply chain disruptions.

6.4. Mergers and Acquisitions in Distribution

The roofing industry is constantly changing. Mergers and acquisitions (M & A) in distribution have resulted in a transformation of the industry. Large companies merge and acquire smaller companies in what was once highly fragmented segment of the industry.

There are several reasons why distribution companies merge and acquire new companies, some of which are to:

- Expand distribution networks
- Broaden product profiles
- Enter new geographic markets
- Strengthen presence in existing markets
- Acquire new talent and expertise

6.5. Roofing Distribution and Sales

Sales in the roofing industry are through two main channels:

1. Specialty distributors, who purchase in bulk from manufacturers and resell to end users, such as roofing contractors.
2. Direct sales, where the manufacturer sells directly to end-users, such as homebuilders, roofing contractors or manufactured housing providers.

Other channels include mass merchandisers and lumberyards. Lumberyards are especially useful for contractors and homeowners in rural and semi-rural areas and are serviced by "two step distributors" who typically sell all their products to lumberyards.

Three distributors together account for 75% of the total sales through distribution in the U.S. roofing market. These are: ABC Supply, Beacon Building Products and SRS Distribution. Many small regional distributors make up the remaining 25% of sales. A specialized roofing distributor can purchase in bulk (which means getting better pricing) and resell in desired quantities.

For example, a distributor may sell a few pallets of shingles to a contractor repairing a damaged house. Those distributors can also use their purchasing power to maintain low prices and quickly react to local market conditions, ensuring prompt delivery of products. Distributors often have nationwide networks of sales locations and warehouses that enable contractors to procure materials quickly. They function as "one-stop" shops and house different products and accessories (sometimes even outside the scope of roofing, such as siding and exterior trim).

Distributors also provide services that can enhance contractor productivity, such as technical and installation advice, marketing assistance, credit programs, app-based and online ordering programs and on-site delivery. Some distributors also offer specialized technology such as drone-based roof measurement systems that can boost profitability.

Leading roof distributors have achieved their market positions in the U.S. by offering a broad range of products (both steep and low slope), opening new locations, acquiring smaller competitors, providing high-value services and offering preferred contractor programs to boost customer loyalty. Figure 6.5.1 shows the branch locations of these distributors, which indicate their market presence. Notice that regions of high population density have multiple locations.

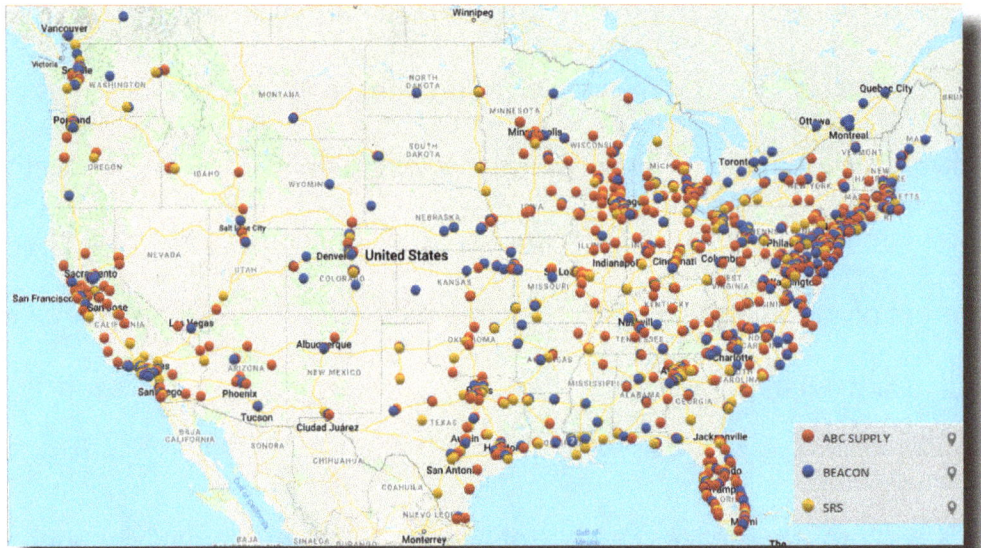

Figure 6.5.1 – Distributor locations
Source: Beacon Building Products

In conclusion, distribution can be defined as the selling and delivering of products and services to customers/end-users. They act as the middleman and control the flow of products from manufacturing plants to different job sites. While it is not mandatory to use a distributor, it can be beneficial to opt for this sales channel.